D0410362

Electricity and Magnetism

Kay Davies
and
Wendy Oldfield

Starting Science

Books in the series

Animals
Electricity and Magnetism
Floating and Sinking
Food
Hot and Cold
Information Technology
Light
Skeletons and Movement
Sound and Music
Waste
Water
Weather

About this book

Electricity and Magnetism introduces children to these forces and how we harness them for our use. Special attention is given to the dangers of using mains electricity and investigations with electricity are always restricted to using batteries. The children learn how to make a circuit and further activities are based on this simple procedure. The invisible power of magnetism is investigated and also how electricity can be used to make a magnet.

This book provides an introduction to methods in scientific enquiry and recording. The activities and investigations are designed to be straightforward but fun, and flexible according to the abilities of the children.

The main picture and its commentary may be taken as an introduction to the topic or as a focal point for further discussion. Each chapter can form a basis for extended topic work.

Teachers will find that in using this book, they are reinforcing the other core subjects of language and mathematics. Through its topic approach ***Electricity and Magnetism*** covers aspects of the National Science Curriculum for key stage 1 (levels 1 to 3), for the following Attainment Targets: Exploration of science (AT 1), Types and uses of materials (AT 6), Earth and atmosphere (AT 9), Forces (AT 10), Electricity and magnetism (AT 11), Energy (AT 13) and Using light and electromagnetic radiation (AT 15).

First published in 1991 by
Wayland (Publishers) Ltd
61 Western Road, Hove
East Sussex, BN3 1JD, England

Typeset by Nicola Taylor, Wayland
Printed in Italy by
Rotolito Lombarda S.p.A., Milan
Bound in Belgium by Casterman S.A.

British Library Cataloguing in Publication Data
Oldfield, Wendy
Electricity and magnetism.
1. Electricity. Magnetism
I. Title II. Davies, Kay *1946–* III. Series
537

ISBN 1 85210 994 7

Editor: Cally Chambers

CONTENTS

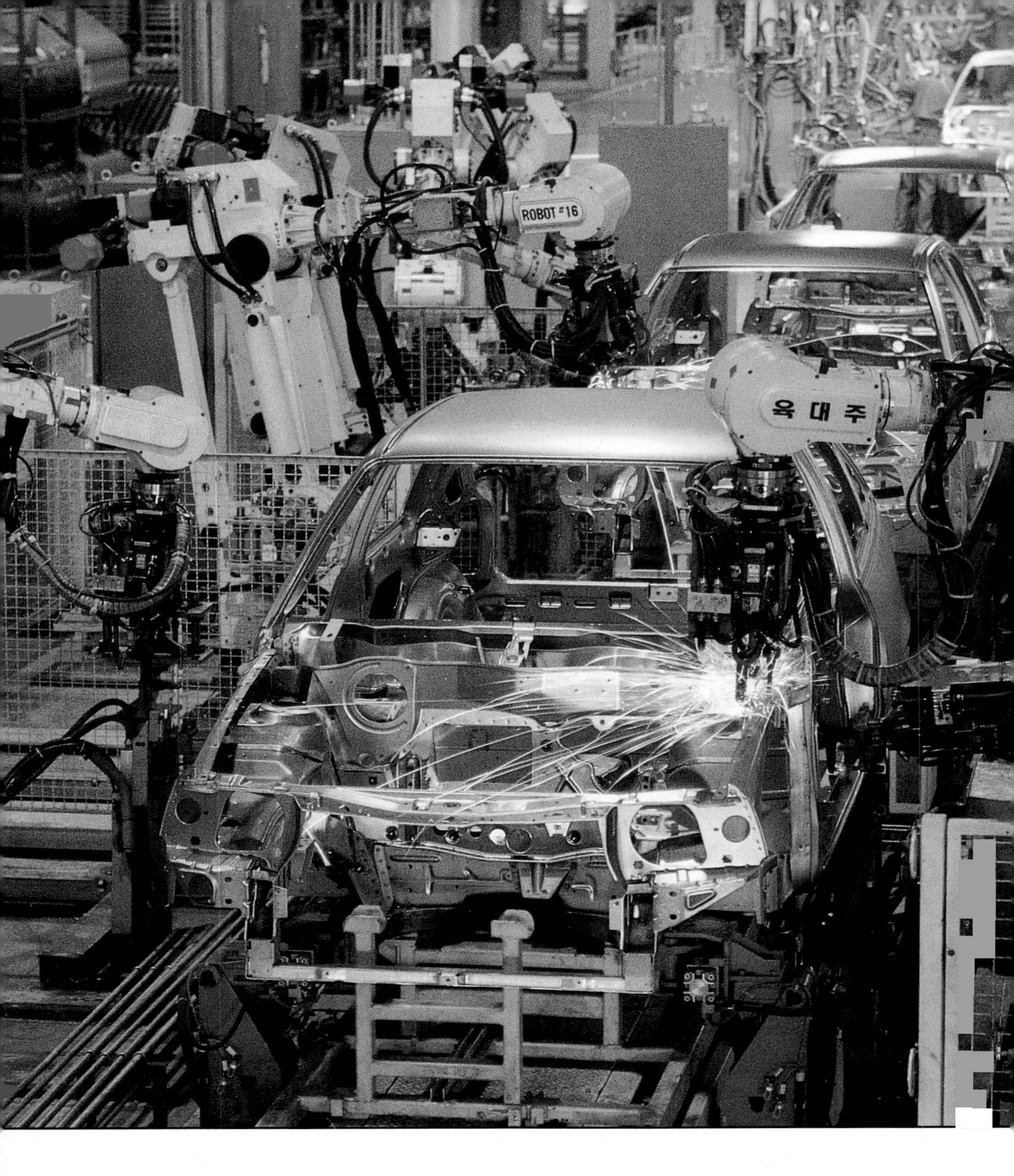

The robots fix the parts of the cars together.
They are worked by the power of electricity.

DRIVING FORCE

We cannot see electricity but it can be very powerful.

Mains electricity is made in power-stations.
It flows as an electric current along wires into our buildings and machines.

We can use it to make light, heat, sound and movement in our homes and at work. It is a useful kind of energy.

How many electrical things can you find in this room?

Make a chart to show if the electricity makes light, heat, sound or movement. It may do more than one.

Electrical item	Light	Heat	Sound	Movement
T.V.	✓		✓	
Iron		✓		

Pylons carry electricity from power-stations to our towns. The wires are above the ground, safely out of reach.

DANGER – DON'T TOUCH!

Many homes have lots of electrical machines.
Each machine has a wire, a plug and a switch.

When we plug into an electrical socket on the wall, the machine is ready to be used. We just have to switch it on.

Mains electricity is dangerous. Never play with plugs, switches, sockets or wires. You could get a painful electric shock. You could even kill yourself.

In the first picture there are six dangerous ways to use electricity. Can you find out what they are?

The second picture shows you how it should be used.

BATTERY-DRIVEN

Batteries don't need mains electricity from the power-station. They have their own supply.

They make electricity when we want to use it. So we can take them with us wherever we go.

A battery can be as small as your thumbnail or bigger than a shoebox.

Batteries are safe to use but they don't last forever.

When the power is used up they have to be changed or recharged.

These all use batteries. Can you think of any more?

Electricity makes the toy robot move along.
The electric power is made in a battery inside it.

Soon the race will start. The cars will keep going around the whole circuit and finish where they started.

RETURN TO SENDER

Electricity must have a complete pathway to travel around. We call this pathway a circuit.

MAKE A LIGHT CIRCUIT

Get a battery, a light bulb in a holder and two wires with crocodile clips at each end.

Clip one end of each wire to the battery.
Clip the other end of each wire to the bulb holder.

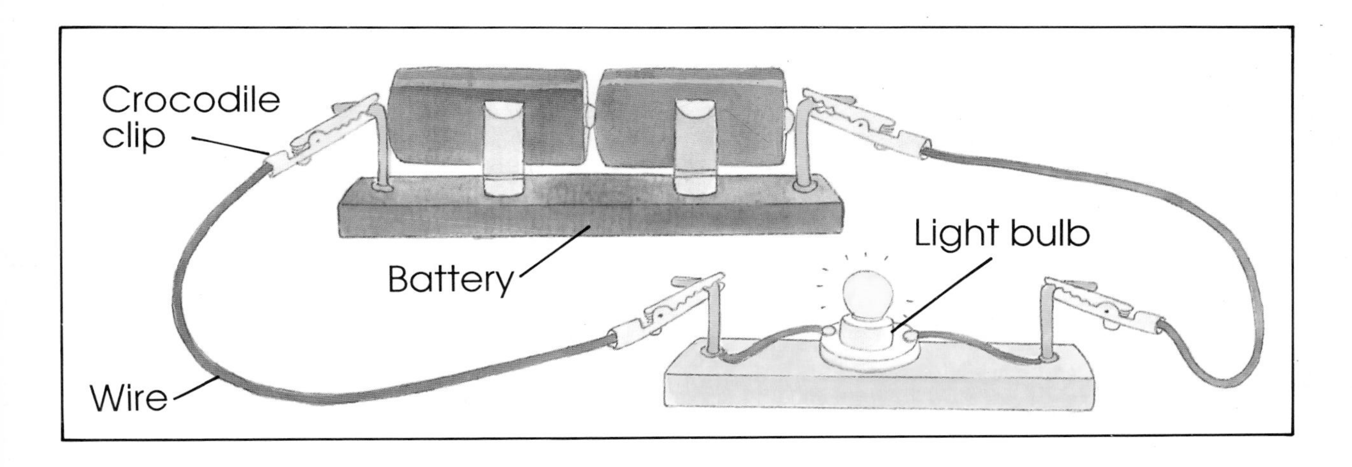

The electric current flows from the battery and down the wire. It passes through the bulb and lights it.

The electricity carries on flowing back to the battery along the other wire.

If part of the circuit is not joined up, the electricity cannot flow at all. Unclip one wire to see for yourself.

The children are having fun on the fairground ride.
Switches make the lights flash on and off.

MAKE AND BREAK

Whenever we use mains electricity we need a switch to turn it on or off.
When we switch on, we complete the circuit. Electricity flows around.
When we switch off, the circuit is broken. The electricity stops flowing around.

USE A SWITCH IN A CIRCUIT

Make a room out of a shoebox. Cut holes for windows and doors. Make a hole in the ceiling for a bulb to fit in.

Clip two wires to the bulb holder. Rest the holder on the roof with the light inside. Finish the circuit like this with a battery, a switch and another wire.

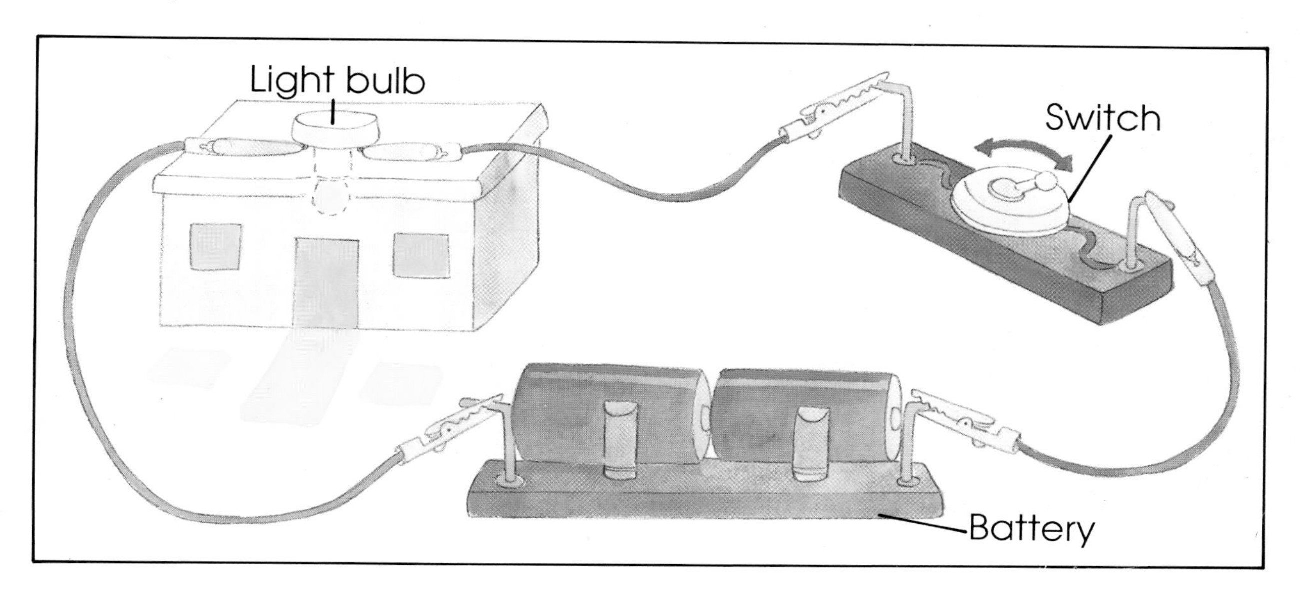

Turn the light on and off with your switch.

PASSING THROUGH

Electricity can flow through some materials. But it cannot flow through others.

Materials which carry electricity are called conductors. The metal in an electric wire or cable is a conductor.

TESTING CONDUCTORS

Collect things made from different materials like metal, wood, card and plastic. Set up your circuit like this.

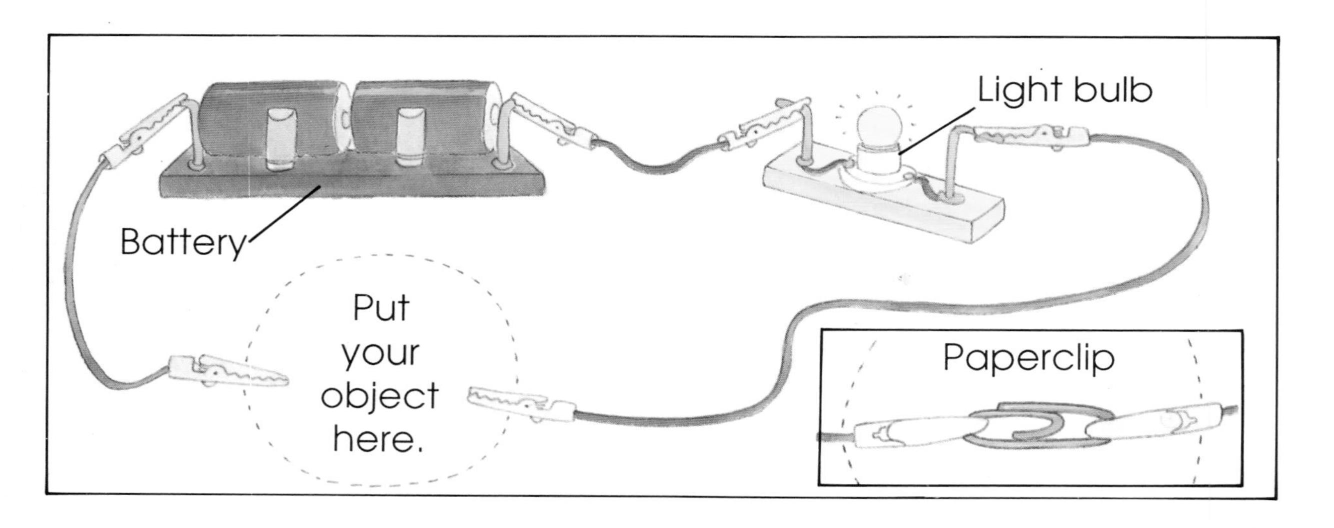

Touch the crocodile clips together to test your bulb.

Then put each object into your circuit in turn, by fastening a crocodile clip to each end of it.

Some conduct electricity to complete the circuit and light the bulb, and some don't. List them separately.

The train speeds through the countryside.
Electricity from the cables above makes it move.

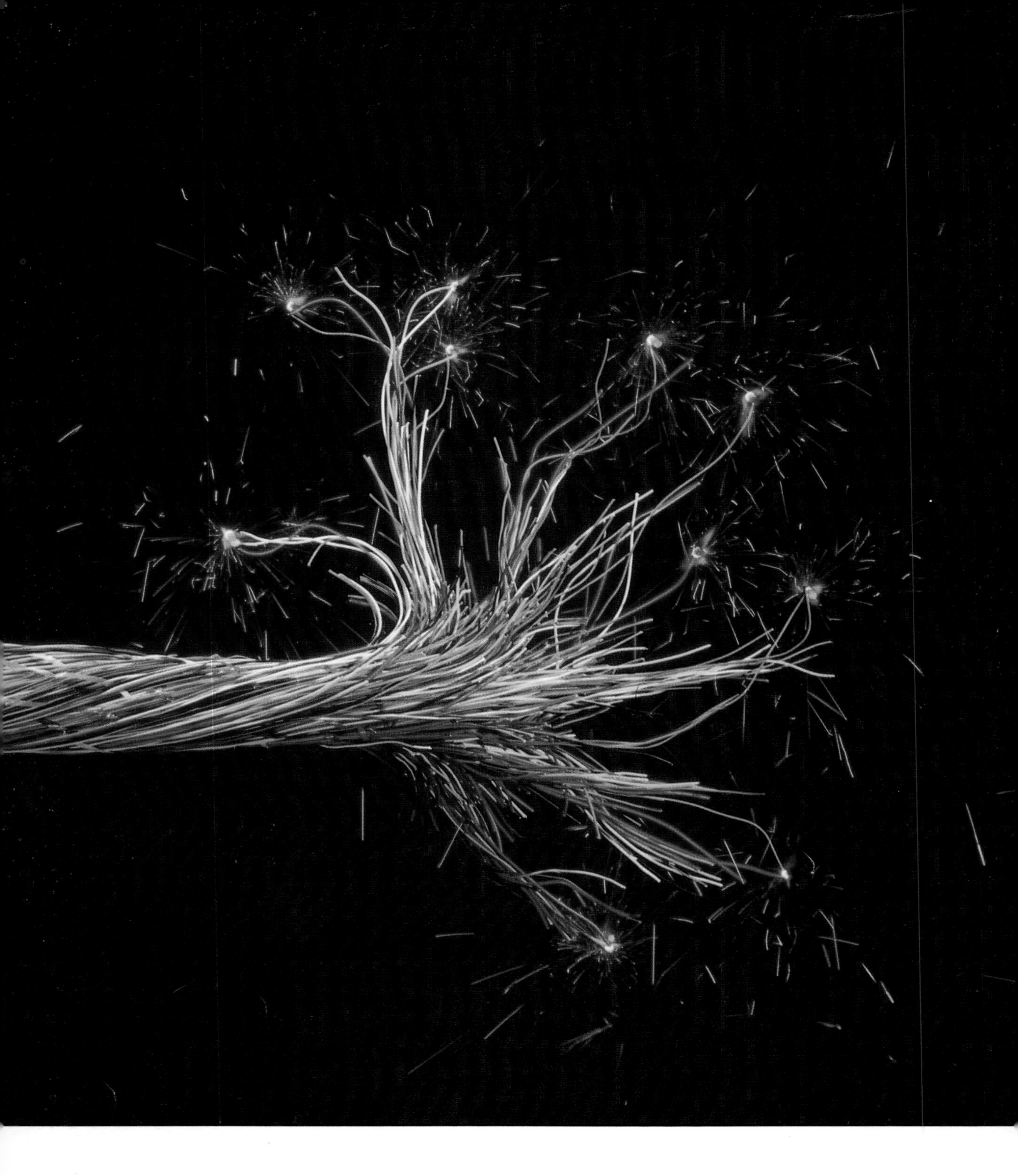

It is very dangerous if wires touch when the insulator is taken away. Never play with wires or cables.

SHORT CIRCUIT

Things that do not conduct electricity are called insulators. We use insulators to cover wires. This keeps the wires from touching and makes them safe.

TESTING AN INSULATOR

Make this 'who lives where' card about 30 cm square. Cut three strips of kitchen foil 40 cm long.

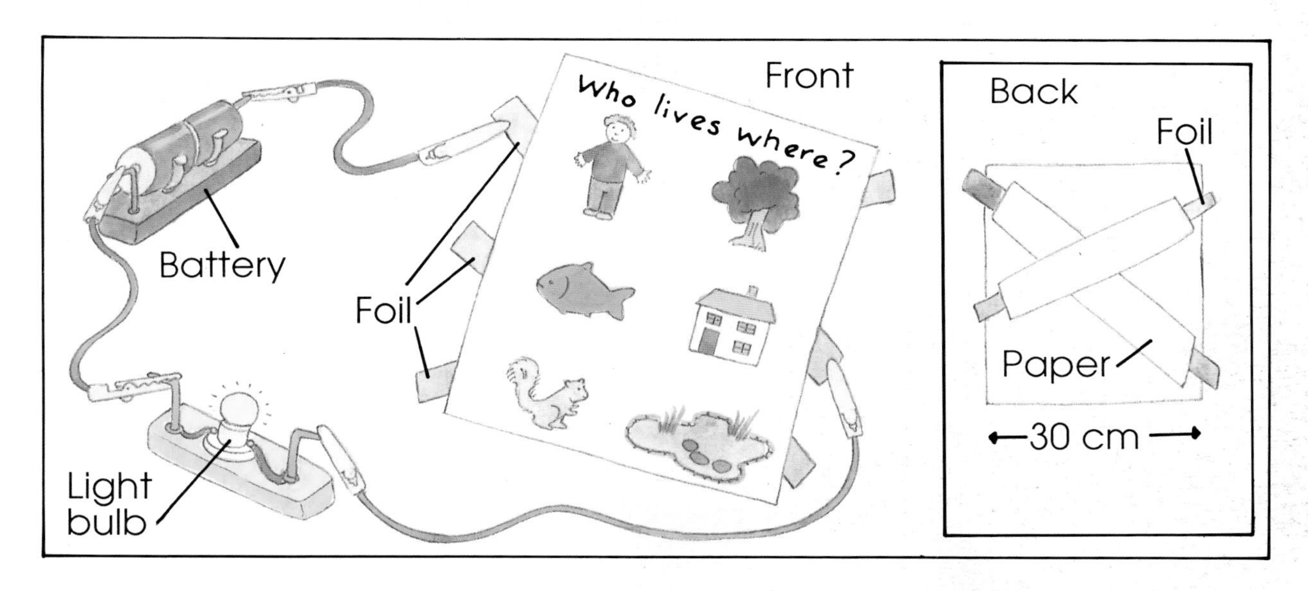

Stick one piece of foil on the back from the squirrel to its tree. Leave some sticking out on each side.

Cover the foil with paper for insulation. Match up the other pictures with foil and insulate them too.

Put the card in a circuit. When you match the pictures correctly, the light will come on.

SHOCKING EXPERIENCE

Lightning is caused by static electricity.

Static electricity can build up when certain things rub against each other.

The electricity passes between two objects and makes a flash.

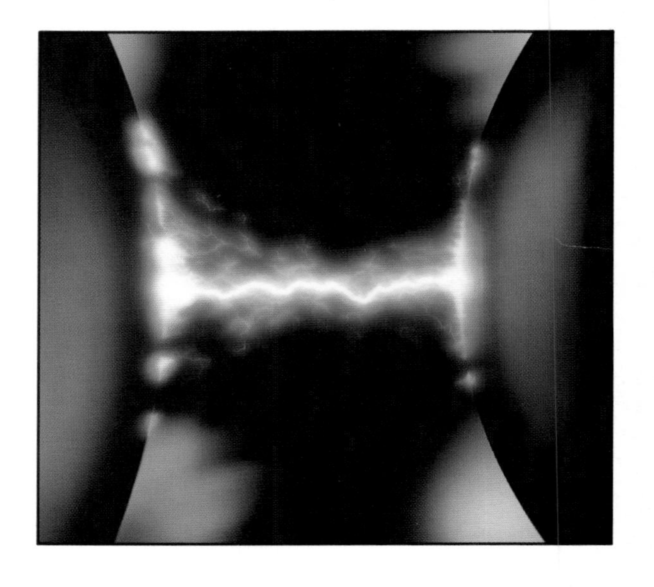

Rub a balloon on your sleeve. Will it stick to the wall?

Comb your hair with a plastic comb. What happens to your hair?

Put the comb near some paper scraps. What happens to them?

Make a collection of plastic, wood and metal objects.

Rub them with a woollen cloth.
Those that pick up scraps of paper hold static electricity.

A flash of lightning from the clouds lights up the sky.
It has the power to damage buildings and trees.

MAGNET MAGIC

Magnets are made of metal which is usually iron. They can be shaped like a horseshoe or as a bar.

Magnets can pull some materials towards them. This is called attraction.

Find lots of objects made of different things.

Use a magnet to find which ones are attracted to it.
What are they made of?

Draw a chart to show which objects were attracted to the magnet.

What happens when you put the ends of two bar magnets together?

Turn them the other way round. Does the same thing happen?

Now turn one round. What happens this time?

Can you feel anything when you hold the magnets together?

Travel games are fun to play with. Tiny magnets inside the pieces stick them to the boards like magic.

STRONG ATTRACTION

Our earth is like a giant magnet. It pulls the needle in a compass to point along a North – South line. This is because the needle in a compass is magnetic too.

You can magnetize your own needle like this.

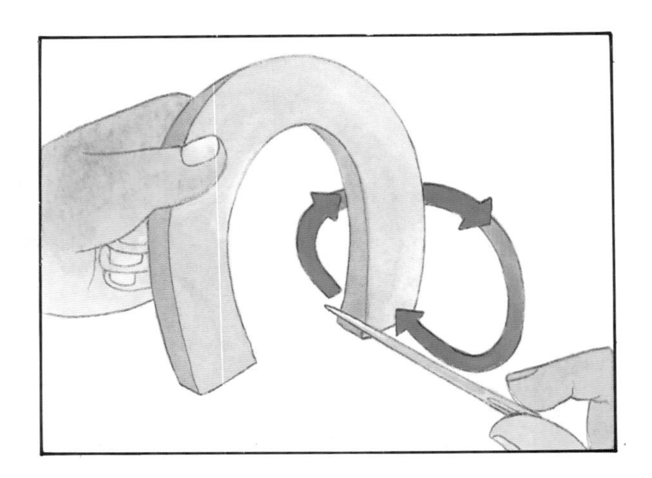

Gently stroke a needle many times with one end of a magnet.

Always move the magnet the same way from one end of the needle to the other.

The needle will now attract a pin. It is magnetized.

Tie a piece of cotton around the middle of the needle.

Hang it from a pencil inside a jar.

When it stops swinging it will point North – South.

Test this with a compass.

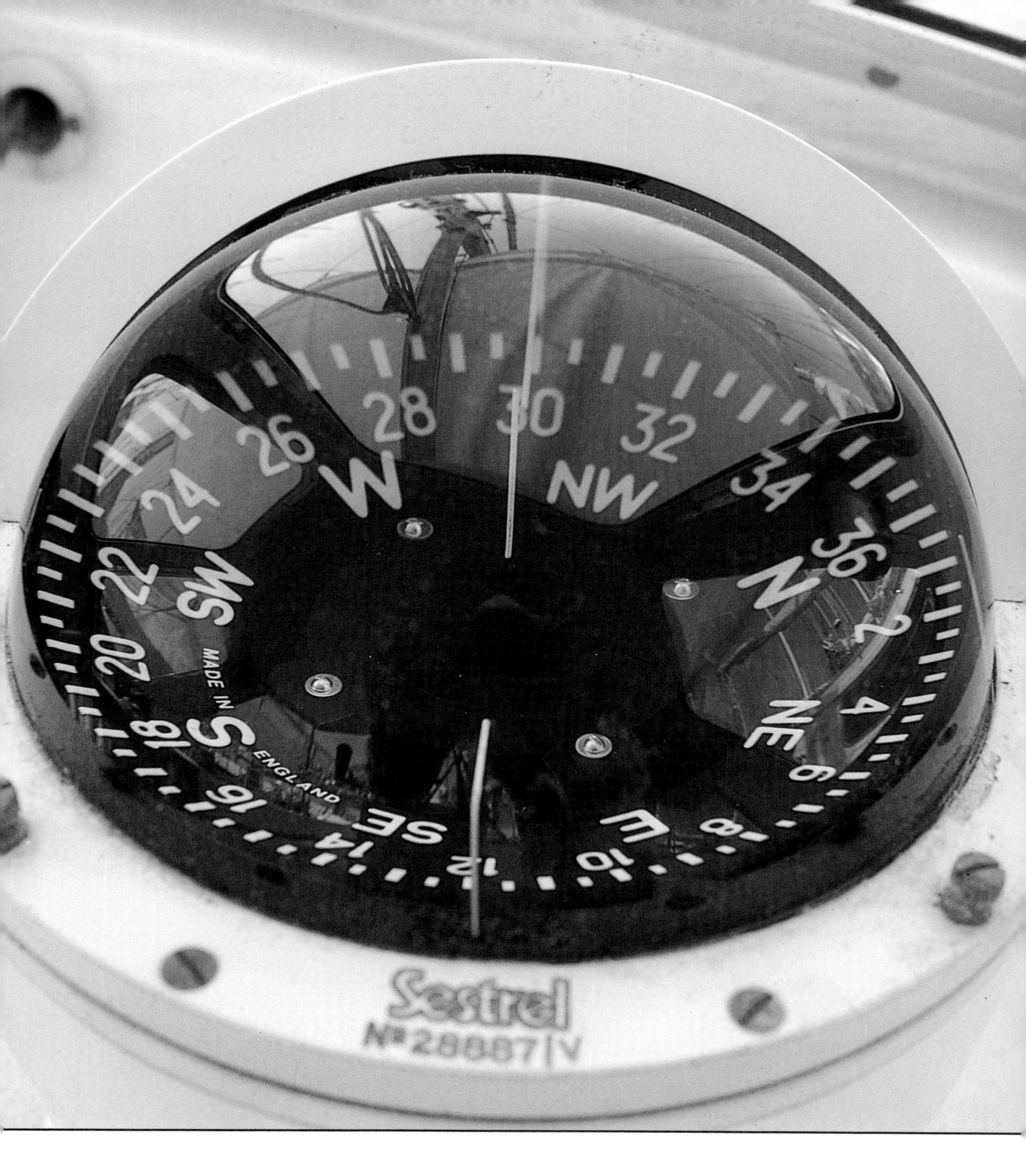

The ship's compass is magnetic. It always points North.
It helps the sailors to find their way.

Magnets inside the objects pin messages and pictures to the fridge. The attraction still works through paper.

POWERFUL PULL

Magnets can sometimes attract through things.
The force field passes through these materials.

Cut a butterfly shape out of card and decorate it.

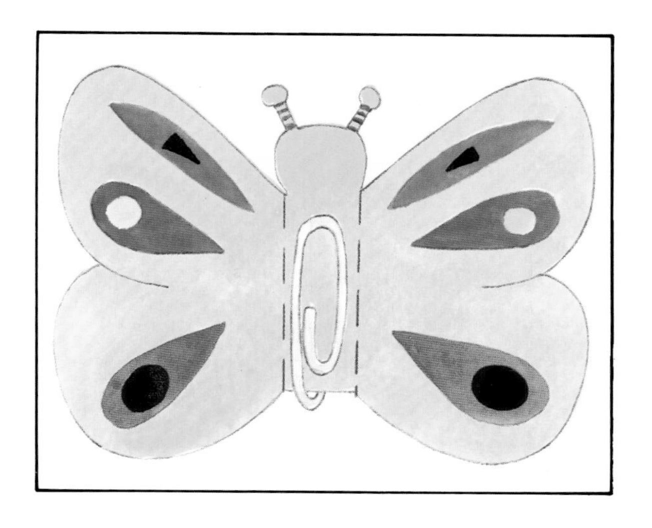

Slide a large metal paperclip along the body like this.

Fold the wings along the dotted lines.

Put your butterfly on wood, glass, plastic, paper and a tin lid.

Move your magnet around underneath.

Can you make your butterfly walk each time?

Try another magnet game with a paperclip. Tie a piece of cotton to a paperclip. Hold the end of the cotton.

Can you make your paperclip rise into the air?
Use your magnet, but don't let it touch the paperclip.

FORCE FIELDS

Magnets are surrounded by invisible force fields. The force fields are strongest at the magnets' poles or ends.

Cover a magnet with a piece of paper.

Sprinkle iron filings over it. Tap the paper gently.

The iron filings show you the shape of the magnetic field.

They show you where the field is strongest too.

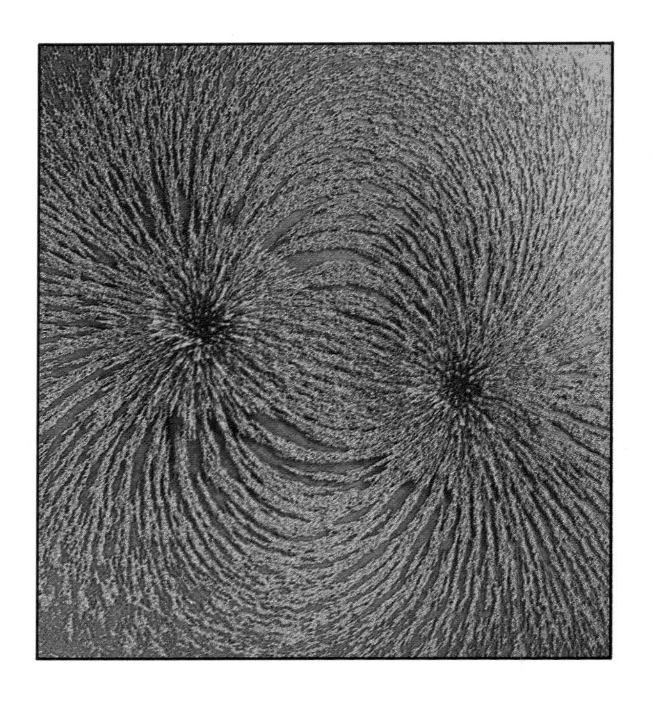

Find lots of magnets to test their strengths.

Try to pick up a string of pins or paperclips with each magnet.

How many will the ends of each magnet hold?

The horseshoe magnet's power pulls the iron filings towards it. They make a pattern on the clear plastic.

CURRENT ATTRACTION

Electromagnets are special magnets. They only work when an electric current flows around them.

Large ones can lift huge weights of iron. If the current is switched off, the iron drops off again.

USE ELECTRICITY TO MAKE A MAGNETIC FIELD

Make this circuit and put a compass on the table. Watch the compass when you bring a magnet near it.

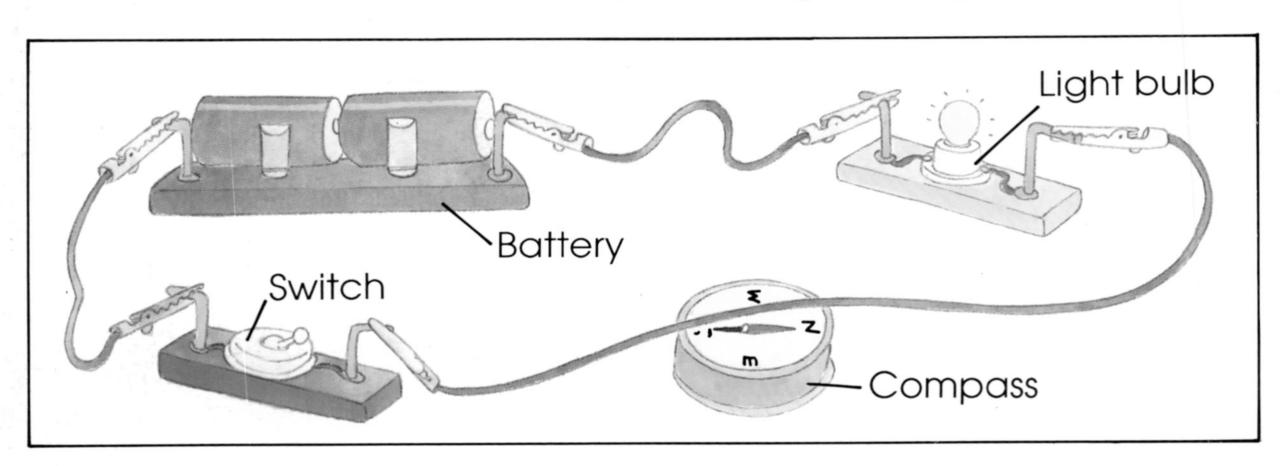

Now hold one of the wires over the compass. Switch the electricity on. Watch what happens to the needle.

Switch off and turn your battery round. Try the test again. What happens to the needle this time?

When the electricity flows it makes a magnetic field. The magnetic field makes the compass needle move.

The electromagnet is fitted to the crane.
It picks up iron in the scrap yard.

GLOSSARY

Battery A source of electric current.

Cables Long hollow tubes often made of plastic. They keep electrical wires safely insulated inside.

Circuit The path followed by an electric current. It must be complete for electricity to flow around it.

Compass A magnetic needle in a glass case. The needle swings to point along a North – South line.

Conductor A material that will let electricity flow through it.

Current The flow of electricity.

Energy Something we have to have, in order to do anything at all.

Insulator A material that will not let electricity flow through it.

Magnetic field The area of attraction around a magnet.

Mains electricity Electricity that is made by power-stations.

Plugs These are attached by wires to electrical machines. Their pins fit into electrical sockets.

Poles The ends of a magnet where the force is strongest. There are North and South poles.

Power The energy to make things do anything.

Recharge To put a store of electricity back into a battery.

Sockets Electrical power points with openings in which to fit electric plugs or bulbs.

Static electricity An electrical charge that builds up on something, caused by rubbing.

FINDING OUT MORE

Books to read:

Finding Out About Things At Home by E Humberstone (Usborne, 1981)
Magnets by Ed Catherall (Wayland, 1982)
Magnets and Electricity by J & D Paull (Ladybird, 1982)

Resources for teachers:

Circuit Training – Junior Education Magazine (November, 1985). Available from: Scholastic Publications, Westfield Road, Southam, Leamington Spa, Warwickshire, CV33 QJH.

Electricity and Magnetism – Junior Resource Pack. Available from: The Education Department, The Science Museum, South Kensington, London SW7.

Energy in Primary Science (Unit 3) produced by the Department of Energy. Further information from your local education office.

For useful films and videos, and a comprehensive selection of resources, send for the school catalogue from: Understanding Electricity, The Electricity Council, 30 Millbank, London, SW1P 4RD.

PICTURE ACKNOWLEDGEMENTS

Aerofilms Ltd 10; Paul Brierly 26 top; Eye Ubiquitous 8, 12, 21, 24; Chris Fairclough Colour Library 21; J Allan Cash Ltd. 9, 23; Science Photo Library 18 top; Wayland Picture Library (Zul Mukhida) *cover*, 18 bottom, 20, 22, 25, 26 bottom; ZEFA 4, 6, 15, 16, 19, 29.
Artwork illustrations by Rebecca Archer.
The publishers would also like to thank Davigdor Infants' School and Somerhill Road County Primary School, Hove, St Bernadette's First & Middle School and Downs County First School, Brighton, East Sussex, for their kind co-operation.

INDEX

Page numbers in **bold** indicate subjects shown in pictures, but not mentioned in the text on those pages.